The La Brea Tar Pits: The History and Legacy of One of Famous Fossil Sites

By Charles River Editors

An illustration depicting a saber-toothed tiger fighting dire wolves in the tar pits

About Charles River Editors

Charles River Editors is a boutique digital publishing company, specializing in bringing history back to life with educational and engaging books on a wide range of topics. Keep up to date with our new and free offerings with this 5 second sign up on our weekly mailing list, and visit Our Kindle Author Page to see other recently published Kindle titles.

We make these books for you and always want to know our readers' opinions, so we encourage you to leave reviews and look forward to publishing new and exciting titles each week.

Introduction

Daniel Schwen's picture of a bubble in the tar pits

"While crossing the basin, the scouts reported having seen some geysers of tar issuing from the ground like springs; it boils up molten, and the water runs to one side and the tar to the other. The scouts reported that they had come across many of these springs and had seen large swamps of them, enough, they said, to caulk many vessels. We were not so lucky ourselves as to see these tar geysers, much though we wished it; as it was some distance out of the way we were to take, the Governor [Portolá] did not want us to go past them. We christened them Los Volcanes de Brea [the Tar Volcanoes]." – Father Juan Crespi, a Franciscan missionary in California

Even at a distance, the acrid stench of asphalt and sulfur singes the hairs of people's nostrils, and when the blustering winds subside, the potent miasma lingers in the air. To the untrained eye, the La Brea Tar Pits seem to be nothing more than simply pools of thick, viscous black sludge, its obsidian-like surface bestrewn with an assortment of autumn leaves and dirt. Gooey methane bubbles spurt up periodically, shattering the glassy veneer of the grease-black lakes, and the shiny bubbles swell to varying sizes and wiggle from side to side before popping, the sticky collapse almost reminiscent of cracking open a chocolate molten lava cake.

This black sludge might seem rather unremarkable after a few moments, as it appears to just sit there in its idle state, but in fact, the seemingly innocuous bubbles are symptomatic of the treacly dark substance lurking on the bottom of the pit. The pit's contents have spelled the doom for a countless number of creatures both large and small, from legions of insects to mighty mastodons, mammoths, and snarling saber-toothed cats from the Pleistocene Era. Of course, this is what makes the area a natural landmark in the first place, and today the La Brea Tar Pits are considered by many scientists to be among the greatest finds in modern history.

Technically, these lustrous lakes of ink-black, while branded "tar," are in actuality pools of asphalt seeps that have remained in place for several millennia, gushing forth from a natural subterranean petroleum spring underneath the city of Los Angeles known as the "Salt Lake Oil Field." Needless to say, the tar pits are a far cry from the glittering, crystalline ponds cooled by the shade of surrounding palm trees found throughout the City of Angels. Indeed, the pungent reek of asphalt, pulsing methane bubbles, and their hauntingly black surfaces, making it impossible to gauge the true depth of the asphalt abysses, should have seemingly served as clear deterrents to the animals that roamed the vicinity prior to their entrapment. Instead, judging by the treasure trove of bones and remnants that have been uncovered within the pits, the sludge seemed to have figuratively emitted a siren song that no animal, regardless of stature or physical power, could resist.

The disturbing and fascinating implications of the silent death traps, situated in what is now 5801 Wilshire Boulevard, Los Angeles, only further heightens their mystery. Evidence shows that the slow, torturous deaths of many of the creatures who became permanently ensnared in the asphalt quicksand were worsened by passing predators who essentially stumbled upon supper served on a sticky platter. Unfortunate, or rather, clumsy predators sometimes slipped, struggled, and were ultimately swallowed up by the tar pit themselves, creating a macabre, yet natural cycle of death and despair.

Unsurprisingly, the La Brea Tar Pits have also become a wellspring of supernatural legends. According to one such legend, the disembodied, bone-chilling shrieks of a desperate woman, supposedly the La Brea Woman, victim of Los Angeles' oldest cold murder case, can still be heard in the dead of the night. More curious yet, these liquid time capsules are swaddled in another layer of mystique, its fossils not only solving mysterious riddles of a bygone age, but also offering up even more questions that are begging to be answered.

The La Brea Tar Pits: The History and Legacy of One of the World's Most Famous Fossil Sites looks at the geological origins of the area and analyzes the fossil finds from the tar. Along with pictures depicting important people, places, and events, you will learn about the tar pits like never before.

The La Brea Tar Pits: The History and Legacy of One of the World's Most Famous Fossil Sites

About Charles River Editors

Introduction

The Dawn of the Black Quicksand

Possession

The Pleistocene Bones

A Legacy for the Ages

Online Resources

Bibliography

Free Books by Charles River Editors

Discounted Books by Charles River Editors

The Dawn of the Black Quicksand

"The La Brea Tar Pits were a place where the skin of the world broke open to reveal the magic underneath, and the life-sized plaster mammoth sculptures emphasized a very important message: do not [expletive] with the tar." – Greg Van Eekhout, "California Bones"

The Spanish term "*la brea*" translates to "the tar" in English, so the name of the La Brea Tar Pits itself is redundant. This was in a way some stroke of serendipity, as one could argue that its alluring, evidently deadly magnetism and its historical significance are both so great that it warrants its name being repeated.

To truly appreciate the multifaceted significance of this bubbling enigma, it's necessary to understand the origins of these bewildering bodies of frothing asphalt. The only problem is that the tar pits' origins continue to be the topic of scholarly debate.

The most popular premise is that the tar pits' origins date back 5-25 million years, when the land that would one day become Los Angeles lay submerged underwater in a shallow expanse that hosted single-celled organisms similar to marine plankton. When these single-celled organisms eventually shriveled up and wasted away, they sank to the seabed, cumulatively transforming into a dense blanket of sediment.

Due to climate change, continents and bodies of land continued to shift as the watery expanse slowly evaporated. The remnants of the deceased plankton, after their once-watery graves were trapped under tons of sediment, gradually recast themselves as oil and gas deposits. The crater was slowly refilled by rainwater and the thaw of the Ice Age, and sediment continued to stack up in tandem with the spinning hands of time. The crushing weight of the increasing sediment, coupled with the immense mass of the ocean, heated up the sheets of carbon-filled organic matter. Owing to the absence of oxygen, the compressed matter then converted to fossil fuels, namely crude oil, or petroleum.

The actual formation of the La Brea Tar Pits occurred about 35,000-40,000 years ago, once the sea had ebbed. Heated founts of petroleum and crude oil miles upon miles underneath the surface bubbled up through the cracks between the splintered rocks above the Salt Lake Oil Field, resulting from periodical earthquakes stemming from the movement of tectonic plates in the Los Angeles Basin, which served as unrestricted pathways to the surface. The coat of petroleum that had traveled up to the surface eventually evaporated, which left behind puddles of rich, syrup-like asphalt, otherwise known as pitch.

An early 20th century picture of the tar pits with oil derricks in the background

The quiet deceptiveness of the inescapable sludge was rounded out by its misleading ability to sustain the growth of flora. The trees and foliage that sprung forth here were enough to seduce the unsuspecting creatures into taking a gander; the leaves and twigs shed by the surrounding flora were the garnishes to these coal-black stews. After about an average of 17-20 weeks, the rigid corpses of the hapless animals that found themselves trapped in the pits vanished underneath the tar's bubbly surface. The pelts and shells of these creatures melted away, but their bones, on the other hand, remained intact, almost expertly preserved by the contents of these sticky time capsules.

The aforementioned process briefly outlines the general consensus on the origins and formation of the La Brea Tar Pits, as well as a backstory for its collection of fossils. Modern scientists from varying schools of thought, however, have spotted holes in the accepted process and put forth their own alternative theories. One of the most telling inconsistencies of the theory, as maintained by these scientists, involves the sizes of some of these tar pits.

To recap the most widely-accepted theory, the pits were birthed from the methane gas and oil that oozed out from the veins of the fractured rock underground, accumulating in craters of different sizes referred to as "blow-holes." On average, those on the larger end measured roughly 15 feet across, with the largest at one point boasting 30 feet in diameter, but fossils were also discovered in much smaller craters of pitch, some of them no larger than a rain puddle, which alternative theorists have claimed contradicts the "tar entrapment theory." La Brea's Pit 36, for instance, measured a mere four feet in length, two feet in width, and had a depth of 11 feet. The six fossils of respectably large carnivores unearthed here (bear in mind, the average length of a saber-toothed cat was anywhere between 5-7 feet), when juxtaposed with the actual size of the pit, rendered such an event highly implausible.

Scientists who subscribe to uniformitarianism and reject the tar entrapment theory have concluded that the bones of La Brea's beasts were transported episodically over the years, carried over in currents from mighty floods that shaped the topography of the region at the foot of the Santa Monica Mountains. Creationists, while also dismissive of the tar entrapment theory, have arrived at a separate conclusion, asserting that all the Earth's creatures were swept away and perished in the Great Flood (estimated to have transpired in 2348 BCE), and that they had drowned before their remnants were aquatically deposited at random into the tar pit craters at La Brea. More earthquakes, as well as the gradual ascension of the Santa Monica Mountain Range, would have also changed the course of certain rivers and the heights of various landforms, leading to more floods. Melted ice glaciers, another catalyst for severe flooding, may have also contributed to the transportation of animal remnants.

The relationship between the La Brea Tar Pits and the early humans who camped out in the area is relatively obscure at best. Apart from the fact that ancient tribes, though lacking a name for "fossil fuels" and "asphalt," were well-aware of their existence and even found many applicable uses of the raw materials, little else is known about notable events that would have taken place in the region until the final years of the Middle Ages. This hazy gap of information only adds to the mystery of these mystical tar pits.

Long before the arrival of the European explorers and settlers, the tar pits functioned as self-replenishing reservoirs of a goo dubbed "*chapapote*," meaning "tar," or "black top," by the Native American Chumash and Tongva tribes. The Chumash were among one of the largest tribes in the region, once numbering in the tens of thousands scattered along 7,000 square miles of the Californian coast. They were a peaceful people who adhered to traditional hunter-gatherer roles and relied on shell beads for currency, but they also valued equality, electing both male and female chieftains. The self-reliant Chumash tribesmen were incredibly resourceful, taking full advantage of their unhampered access to both the land and the sea. Their creativity was, for one, exhibited in the construction of their homes, which could accommodate up to 50 at one time. These were sprawling mounds fashioned out of meticulously bent willow poles, crosswise-tied networks of bulrush, cattail plants, and other grasses, with roofs made of tulle, and reed curtains as room partitions. Robust whale bones, a resource from the sea, were often used as reinforcement materials.

The Chumash people were renowned for their fine craftsmanship and efficiency in boat-building, once ranked the best of all the Native American tribes in the state. The tribesmen collected redwood trunks and planks of driftwood floating upon the Santa Barbara Bay, which were then assembled and the cracks of the canoes plugged with gobs of asphalt taken from the tar pits. *Chapopote* served as a panacea for all leaks and reinforcement requirements. The pits' asphalt was also used to waterproof baskets, while hardened tar was also used as roofing material and for kindling. The tribesmen appeared to have been cognizant of the tar pits' lethally adhesive capabilities, and they may have capitalized on the vulnerability of the animals trapped in the pits.

The Chumash's neighbors, the Tongva people, otherwise known as the "Gabrieliño-Tongva" tribe, resided in tracts of land measuring 4,000 square miles by the Los Angeles Basin and the Southern Channel Islands. The territory was christened "*Tovaangar*," meaning "the world," by its residents, the first of the island, who conversed in the Tongva tongue, often said to be the "first language of Los Angeles." Much of the first Spanish settlements and missions owed their existence and productivity to the Tongva people, who were shackled, reduced to slaves, and subjected to debilitating manual labor at the hands of colonizers.

The Tongva tribes, like their Chumash brethren, also availed themselves of the La Brea Tar Pits, primarily using them to caulk their canoes. The Chumash and Tongva's superior, asphalt-strengthened boats allowed them to fish for longer hours and in deeper waters, and these vessels granted them access to other resources otherwise unavailable on land. The tribes' blossoming wealth in resources, in turn, turned them into expert tradesmen who bartered and swapped valuable items along the coast.

Later generations of the Chumash and Tongva tribes continued to use *chapapote* as sealants, and *chapapote* remained essential for every repairman. La Brea tar was used to glue together flutes, bone whistles, musical rattles, and other instruments, as well as abalone dishes, pipe mouthpieces, and containers made out of shells. *Chapapote* even penetrated tribal fashion, as women smoothed the blades of grass on their skirts with small hunks of tar. Superglue and waterproofing aside, the tar was also used in the casting of broken bones, and as chewing gum.

In fact, modern historians have reason to believe that the tribes may have been far too well-acquainted with the sticky substance, which may have been rather detrimental to their health. In a study published in the *Environmental Health Perspectives* journal in May of 2011, researchers examining the skeletal remains of 269 Chumash men and women discovered a gradual, but pronounced decrease in the sizes of their skulls over a period of approximately 7,500 years. The poison of the *chapapote* comes from the polycyclic aromatic hydrocarbons (PAHs) that natural asphalt is laced with, which also activates when heated. The noxious fumes inhaled by the tribesmen may have also contributed to their shrinking skulls.

Along with their diminishing skulls, the heights and physical builds of the Chumash tribe also experienced a decline with time. Sabrina Sholts, an anthropologist tenured at the University of California, Berkeley, explained, "The decline we're talking about is a very gradual process over thousands of years, and it could have been chemicals, these carcinogens [they] were exposed to everyday – the equivalent of smoking and breathing polluted air." Another relevant study demonstrated how effortlessly the PAHs of bitumen tar are able to infiltrate the body – passable via inhalation, ingestion, and skin contact – and how easily the naturally occurring chemicals can spread and taint various organs, as well as fetuses in wombs.

Apart from these tribes, the La Brea Tar Pits remained seemingly untouched and undiscovered until the summer of 1769, when a Franciscan friar by the name of Juan Crespi, who was tasked with chronicling the expedition of future Spanish Governor of the Californias, Gaspar de Portola, made the first recorded reference to the "springs of pitch" in La Brea.

Crespi, born on the Spanish Majorca island in 1721, was little older than a teenager when he entered the Roman Catholic Franciscan brotherhood. Crespi sought knowledge in addition to a connection to a higher power, and enrolled himself at Palma's Lullian University, where he forged a strong academic and personal relationship with his philosophy professor, Junipero Serra. The Petra-born Serra, also a Franciscan friar who would later go on to found at least 10 Catholic missions across California, was dispatched to New Spain, Mexico for missionary work in 1749; the pious, well-schooled, and adventurous Crespi tagged along. Here, Crespi assisted Serra in the evangelization of native folk and settlers alike, and again joined Serra when, in 1767, he was sent to Baja in southern California. Two years later, Serra was handpicked by de Portola to proceed to Alta California, where he was expected to erect a number of evangelical colonies. His trusted associate, Crespi, was yet again charged with assisting in the establishment of said settlements, as well as documenting the so-called "Sacred Expedition."

Portola, Serra, Crespi, and 67 others ventured north in the early summer of that year. On the 1st of August, the procession of expeditioners wound up in what is now Los Angeles, or more precisely, a Tongva village called "Yang-na." It took two days for Crespi's wandering eyes to spy the curious-smelling tar pits.

An excerpt from Crespi's journal entry, dated August 3rd, 1769, reads, "We proceeded for three hours on a good road; to the right were extensive swamps of bitumen which [the tribes] called 'chapapote.' We debated whether this substance, which flows melted from underneath the earth, could occasion so many earthquakes." Not only was the friar the first known Western explorer to chance upon the tar pits, he was the first to label the pools of asphalt as "bitumen."

Crespi became convinced that the Tongva territory was once upon a time inhabited by a winding range of volcanoes, even referring to tar mounds as "tar volcanoes," and he suggested a connection between the tar pits and the area's propensity for earthquakes. The excerpt from Crespi's journal entry continued, "This afternoon, we felt new earthquakes, the continuation of which astonishes us. We judge that in the mountains that run to the west in front of us there are some volcanoes, for there are many signs on the road which stretches between the Prociuncula River and the Spring of the Alders, for the explorers saw some large marshes of a certain substance like pitch; they were boiling and bubbling, and the pitch came out mixed with an abundance of water. They noticed that the water runs to one side and the pitch to the other, and that there is such an abundance of it that it would serve to caulk many ships." Crespi did not yet know it, but his was a groundbreaking discovery: the first clear sign of the presence of petroleum in the western half of North America.

The influx of European settlers in the decades that followed resulted in many more travelers crossing paths with La Brea's cluster of tar pits. One such passerby, a Spanish naturalist named Jose Longinos Martinez who opted to go solo after an irremediable squabble with his colleagues from a botanical exhibition, made the following note of the tar pits in his diary: "Near the Pueblo de Los Angeles there are more than 20 springs of liquid petroleum, pitch, etc. Farther to the west of said town...there is a great lake of pitch, with many pools in which bubbles or blisters are constantly forming and exploding. They are shaped like conical bells and, when they burst at their apex, they make a little report. I [attempted to examine] the holes left by the bubbles, but when they explode they are followed by others in succession and gave one no opportunity to examine the cavity."

Jedediah Smith, the renowned frontiersman and explorer of the Rocky Mountains, was accompanied by Harrison G. Smith when they were presented with a small lump of the hardened *chapapote* during their stop at Mission San Gabriel in 1826. Smith noted that the solidified variant was used by the "citizens of the country...to pitch the roofs of their houses."

Unbeknownst to these oblivious explorers, they had just stumbled upon an unlikely, yet fantastic nest of historical jewels that would aid in illustrating the lives of prehistoric creatures, thereby supplying the answers to some of nature's most evasive mysteries.

Possession

"I am the tarred and feathered stork

Who flapped its limbs until they stuck...

This is the asphalt killing-ground,

A lake that thirsts. Beware. Be warned..." - Duncan Forbes, "La Brea"

In 1828, a 4,439-acre (roughly seven square miles) strip of land encompassing the tar pits was designated the title of "Rancho La Brea" via a Mexican land grant, and it was entrusted in the care of a pair of business partners named Antonio Jose Rocha and Nemisio Dominguez. The grant, which was authorized by the three-term alcalde (mayor of a Spanish town) of Los Angeles, Jose Antonio Carrillo, came with the stipulation that the tar pits remain public property. The proviso allowed registered residents of the surrounding pueblo to dip into the communal tar pits as they so pleased. Most used the asphalt to reinforce the roofs of their adobe dwellings. The deed to Rancho La Brea, which comprised of sections of what is now West Hollywood, present-day Hollywood's Wilshire's Miracle Mile, was formally confirmed by the governor of Alta California, Jose Maria de Echeandia, and was publicly acknowledged by Governor Juan Alvarado again in 1840.

Given Rocha's impressive background, the accomplished entrepreneur seemed naturally well-suited to preside over the prized estate. Not only was the Portuguese-born Rocha hailed as being a "pious man, a favorite will all the priests," the résumé of this competent jack of all trades included stints as a carpenter, blacksmith, weaver, sailor, gunsmith, and soldier. The business-minded investor is also remembered fondly as the first official Portuguese settler of California, and for constructing a strapping adobe structure, the Rocha Adobe, upon a 250-acre plot of land within the ranch. The building was rebranded as Los Angeles' first City-County Municipal Building in later years.

That said, even the perceptive Rocha was blind to the cradle of scientific knowledge present in the area, as well as the ripples it would potentially cause in the scientific world, even as it all lay in what was effectively his own backyard. Rocha later received his business partner Dominguez's relinquished shares for unspecified reasons.

Acquiring the ranch was undoubtedly a judicious business decision, but the ever-changing tides of politics came with throbbing headaches that its proprietors were forced to tackle. Following the incorporation of California into the United States as a consequence of the Mexican-American War, the issue of land ownership swiftly became a disputatious and thorny subject. As per the terms of the 1848 Treaty of Guadalupe Hidalgo, which yielded the United States an additional 525,000 square miles of territory, including California, large swathes of Colorado, Utah, Nevada, Arizona, and half of New Mexico, all existing land grants were presumed to still be valid. Just three years later, however, Congress passed the California Land Act of 1851, which was established with the intention of re-investigating the legitimacy of existing land grants. As such, all landowners, including Rocha, were expected to substantiate ownership of their lands in court.

In the spirit of due diligence, Rocha, along with his son Jose Jorge Rocha and Josefa de la Merced de Jordan, jointly lodged a claim with the Public Land Commission the following year. Much to their dismay, the case of Rancho La Brea remained in limbo for the next eight years, until it was repudiated by the court in 1860. Even so, the ranch owners persistently petitioned for the court to retry their case in a series of hearings, and they enlisted the aid of a former major, lawyer, and surveyor of the city of Los Angeles, Henry Hancock, to spearhead their efforts.

Hancock

    As the *rancho* team battled it out at court, scientists across the state began to take steps to assuage the scientific interest that the La Brea Tar Pits had piqued. In 1853, shortly after New York-born geologist William Phipps Blake was appointed lead geologist of the Pacific Railroad Survey of the Far West – chiefly to analyze, record his observations, and arrive at a theory on erosion in the geologic features of southern California – he paid a visit to the ranch. Blake became the first scientist to truly scrutinize the pits and its bitumen in depth, pinning his focus on a large pool of pitch measuring 30 feet in diameter. He made a marked reference to the bitumen's tendency to spill over and mix with pieces of sediment, which hardened into a thick layer, but maintained a soft consistency in the heart of the pit.

Blake

A snippet from Blake's report explained, "This spring was nothing more than an overflow of the bitumen from a small aperture in the ground around which it had spread on all sides, so that it covered a circular space about 30 feet in diameter. The accumulated bitumen had hardened by exposure and its outer portions were mingled with sand, so that it was not easy to determine its precise limits. It formed a smooth, hard surface like a pavement, but toward the center it was quite soft and semi-fluid, like melted pitch."

The elusive vertical drop of the blend of bitumen and sediment, its body largely cloaked by alluvial topsoil, was later determined to be an estimated one or two feet in depth. Altogether, the firm layers of bitumen and sediment, described as "fields of hardened asphalt and petroleum," whipped up by the ranch's tar pits were estimated to be anywhere between 160-600 acres. Blake and his colleagues were tickled by the peculiar properties of the seemingly solidified pitch, which, all the same, bore fissures that allowed gushes of oil underneath to leak through its cracks.

The proprietors of the ranch were presumably flattered by the growing scientific interest unfolding around the property's tar pits, but they were also disheartened by what seemed to be a losing battle. The stress of it all continued to climb as they burned through their funds, which were mainly used to settle their continuously mounting legal fees. When the Rochas came to terms with the reality that they could no longer shoulder the costs in the summer of 1860, the patriarch's son and business partner, Jose Jorge, unloaded the deed to their lawyer, Hancock, for pennies on the dollar.

In addition to Hancock's surveying skills and counseling prowess, the New Hampshire native and Harvard-trained lawyer was equipped with an enterprising mind. Having run away at the age of 12, whereupon he became financially independent through his earnings as a crew member on a mackerel schooner, the self-made man knew just how to take advantage of different circumstances. Hancock pooled together $20,000 ($526,810 today) from the profits he had derived from a glittering investment in a placer mine and purchased roughly two-thirds of the ranch, including the tar pits from the Rochas, with each acre priced between $2 and $3 apiece. His brother, John Hancock, later acquired small portions of the ranch in the 1860s. The legal victory the Rochas had pined after for so long only came in 1873, when the court finally awarded them the title, addressed to one "A. J. Rocha et al." Of course, by then, the Hancock brothers were deemed the majority owners of Rancho La Brea, with Hancock holding the deed to 2,400 acres and his brother John holding 1,200 acres.

Not long after acquiring his share of the ranch, Hancock immediately went to work on exploring and exploiting the commercial potential of the tar pits on his turf. Hancock established a proper asphalt quarry business, and he sought out and brokered deals with construction companies statewide.

He then went to work marketing the pitch as affordably priced, ethically sourced building material used in the paving of sidewalks and streets. An article published by the *Los Angeles Times* profiling Hancock and his new business succinctly detailed the essentials of the enterprise in the following passage: "Major Hancock had about 25 Chinese [laborers] at work in digging out the best of it. The material was conveyed to large open iron boilers where it was boiled for 24 hours. It was then poured into long trenches in the ground, broken up, and shipped to San Francisco for paving purposes." Hancock's asphalt ore was priced at about $20 (approximately $455 today) for a ton, and the asphalt was also used in the preservation of water pipes and railroad ties.

The Rochas' decision to soldier on in court all those years continued to cost them, even after the fact. The recognition of their deed came with a clause that mandated the Rochas to compensate ex-US Senator Cornelius Cole, who had agreed to appear on their behalf before the Supreme Court. Thus, for his services, Cole was allotted 500 acres, roughly a tenth of the ranch. The former senator's tract of land, which stretched from what is now Sunset Boulevard to Rosewood, and from Gower to Seward, was aptly renamed "Colegrove." Small portions of the land were also divvied up and purchased by businessmen James Thompson, who went on to build the first official residence within the ranch. When Thompson lost his share following a string of poor decisions and the declaration of bankruptcy, Arthur F. Gilmore and his partner Julius Carter purchased Thompson's plot of land and converted the Rocha Adobe to a dairy farm.

A Harvard geology professor, geologist, and expert of economical geology by the name of Josiah Dwight Whitney paid a visit to Rancho La Brea in 1865, in the midst of the Rochas' legal battles. He added to what was then the small, yet promising trove of information regarding the ranch's tar pits with his treatise, published that same year: "Report of Progress and Synopsis of Fieldwork from 1860 to 1864: California Geological Survey, Volume 1." The most noteworthy passage from the report penned by Whitney reads, "Over a space of 15 or 20 acres the bituminous material, which, when seen by us, in the winter, had exactly the consistency and color of tar was oozing out of the ground at numerous points. It hardens on exposure to the air and becomes mixed with sand and dust blown into it, and is then known as 'brea.' The holes through which the bitumen comes to the surface are not very large, few being more than 3 or 4 inches in diameter. On removing the tarry substance from the holes, by repeatedly inserting a stick, the empty cavity was very slowly filled up again...A very large amount of the hardened asphaltum, mixed with sand and the bones of cattle and birds have become entangled in it, lies scattered over the plain."

As that passage suggests, Whitney was under the assumption that the bits of bones found when sifting through the pits belonged solely to careless barn animals and other small critters who had wandered into and became trapped in the sludge. This was not a groundless premise, because the partially engulfed corpses of finches, squirrels, rabbits, raccoons, weasels, and other varmints could be seen peeking out from the surfaces of these tar pits. Even larger animals such as sheep, horses, and cattle were sometimes paralyzed by the bubbling pools of pitch, unable to wrench their legs or ankles from the sludge without assistance. Thus, Whitney surmised that if the wounded neighs and guttural bleats of these trapped barn animals had fallen on deaf ears, they, too, would have perished from either dehydration or starvation. He did not think much of the more sizable petrified remnants plucked from the pits, initially brushing them off as the carcasses of coyotes, deer, and other wild fauna. It simply did not occur to Whitney or his colleagues that these very bones could have been thousands of years old.

During this time, the ranch began to receive a growing litany of complaints from disgruntled residents of the pueblo over the pesky bones and chunks they were made to fish out from their extractions. Like Whitney, the asphalt collectors were ignorant of the historical gems they had unwittingly panned. It took another decade for a different scientist to acknowledge the possibility that these were not the bones of barn animals and wild creatures, but the fossilized remains of prehistoric beasts.

In 1875, William Denton, a British geology professor and leading psychometry specialist based in Boston, was sent to Rancho La Brea to appraise the economical viability of the oil underneath the property. Thus, the prehistoric connections drawn by Denton were themselves an accident of sorts, but it appeared as if Denton had been destined to disinter the remnants' association with the far past because Denton had authored several books that indicated a heartfelt interest in fossils. Denton raved in *Nature's Secrets; or Psychometric Researches*, published in 1863, "From the first dawn of light upon this infant globe, when round its cradle the stormy curtains hung, Nature has been photographing every moment. What a picture gallery is hers!"

It was none other than Henry Hancock who brought a jagged, curved fang of an unidentified carnivore he had recovered from one of the pits to Denton's attention, thereby indirectly triggering the pivotal revelation. Denton spent many hours admiring the chipped canine tooth, which measured roughly 9.5 inches in length and about 3.5 inches across. Upon his conclusion that the tooth would have measured approximately 11 inches in length if intact, it dawned on him that the tooth was far too large to belong to any existing creature. After further investigation, Denton speculated that the tooth belonged to a feline carnivore that stalked the Earth during the Ice Age. This theory itself was fairly controversial at the time, as the *smilodon fatalis* family was only officially recognized by paleontologists in 1842.

By the time Denton was due to depart from the ranch, he had unearthed a handsome medley of bones, including the remains of the Equus genus, a type of extinct horse, or perhaps the stilt-legged *Haringtonhippus*, which he parceled carefully and brought home with him to Massachusetts. Once he had settled in, Denton eagerly composed a report entitled "Proceedings of the Boston Society of Natural History," announcing on his momentous discoveries. Much to his disappointment, the work inadvertently slipped under the radar, sparking little interest at all within the scientific community.

In spite of Hancock's solid business acumen, when he died in 1883 after a private battle with an unspecified illness at the age of 61, his widow, Ida Ross Hancock inherited his 2,400 acres of the ranch along with a mountain of debt. Ross, the daughter of the esteemed Hungarian Count Agostin Haraszthy, a prominent local vinter, had a relatively luxurious upbringing, but unlike many in her shoes, her privilege was balanced by an industrious spirit and a good head on her shoulders. Shortly after Hancock's demise, the heartbroken Ross, who could not bear to be surrounded by the painful memories of her dearly departed husband, moved up north to San Francisco to spend some time with her side of the family. It quickly became clear, however, that it was bordering on impossible to tend to the affairs of Rancho La Brea from a city close to 400 miles away. The impracticality of this system was also exacerbated by a fresh set of problems, the most urgent of which pertained to the illegal squatters who had taken it upon themselves to set up camp within the ranch's borders and were filing hollow, but nonetheless troublesome ownership claims.

Ida Ross Hancock

When the threat of losing her husband's legacy became all too real, Ross cast aside her personal disinclinations, gathered her sons, George Allan and Bertram, and moved back into the ranch. The family bade farewell to the elegant Haraszthy estate and reacquainted themselves with the humble shack the couple had raised within a colorful grove of "eucalyptus, pepper, and palm trees" by the tar pits. It was here that Ross, now affectionately remembered as the "Lady of La Brea," vowed not only to hold down the fort, but to restore the prestige once associated with their family name.

The drive of the surviving Hancock widow was indestructible. Nosy neighbors and other members of the community began to whisper amongst themselves when they spotted Ida and her sons hawking watermelons, beans, and other crops to passersby for extra cash. Meanwhile, she concentrated on revising and revitalizing marketing tactics for their limitless reserve of "brea," as well as searching for other avenues of profit. As she slowly footed the bills, both debt and domestic, she became increasingly reliant on the tar pits as her primary source of income. Soon, she was no longer struggling to cope with her living expenses and family investments, but was instead ahead of her bills, which covered living costs, the salaries of ranch workers, the maintenance of the ranch and its necessary equipment, and taxes.

Ross discovered another way to capitalize on the rivers of black gold that lay underneath the ranch. In 1885, the competent businesswoman scored her first short lease with the directors of Union Oil Company of California (later Unocal), Lyman Stewart, Dan McFarland, and Wallace Hardison. In exchange for the "rights and privilege" to access the tar pits, nicknamed the "Asphaltum Springs," the Hancock family reserved agricultural rights, maintained the right to continue mining the asphalt alongside Union Oil, and received an eighth of the royalties derived from the company's profits throughout the duration of the lease. Union Oil proceeded to drill a total of four wells on the property, but only one managed to tap into the oil reserve with some success. Still, news of the moderate success of oil drilling on the ranch perked the ears of other eavesdropping oil firms. Not long thereafter, she negotiated another lease with a competing oil company under similar terms, which led to an uptick in her annual income.

Throughout the following decade, Ross continued to delineate borders across her property and racked up one leasing agreement after another. The year 1893, though permanently darkened by the death of her son, Bertram, who had succumbed to complications stemming from typhoid fever at the age of 15, came with a minute, but meaningful triumph: her property was renamed the "Ida Hancock Tract" and officially surveyed for the first time, which was said to be confined by "Prospect Avenue on the north, Highland Avenue on the west, Sunset Boulevard on the south, and Seward Street on the east." By 1894, the coveted lots on the tracts were listed on the market as $300 an acre ($9,081 today), a far cry from the $2 her husband had shelled out for them. George Allan, was employed to manage the ranch and received an ample $100 ($3,027 today) each month for his services.

The dawn of the 20th century gifted the Hancock's with their most prodigious and profitable partnership. In 1900, Ross and Hancock, Jr., signed off on a 24-year drilling lease with the Salt Lake Oil Company, which resulted in the designation of the Salt Lake Oil Fields. The lucrative partnership kicked off with a smashing start when the first oil well bored by the firm struck such a dense and powerful pocket of oil that the surging column of black gold blasted the casing clean off the drill. In just under 11 years, the firm drilled a total of 258 wells on the Ida Hancock Tract.

Inspired by the success of his mother's ventures, Hancock, Jr. was determined to forge his own path in life, and he acquired a loan to the tune of $10,000 from his mother (about $257,158 today) to found his own oil firm, which he dubbed "La Brea Oil." Thanks to the mediation skills and entrepreneurial sagacity he had picked up from his parents, La Brea Oil was a capital success. By 1907, the Hancock firm was raking in an average of $1,000 ($23,480 today) in revenue each day, with the sound operation churning out approximately 300 barrels across 70 of the company's own wells. Ida Ross Hancock was eventually crowned the richest woman in all of Los Angeles, and her stellar reputation was bolstered by her generosity and acts of philanthropy.

Decades later, the once-rolling torrents of liquid petroleum swimming beneath the earth began to decelerate in their flow, and many of the oil derricks that once dotted the Ida Hancock Tract have since been torn down in favor of a stunning selection of stately homes in what is now "Hancock Park." In later years, upscale, chic neighborhoods were established by developers who purchased lots in the northern neck of the ranch, inhabited and frequented by stars, starlets, directors, producers, and others who were mainstays of Hollywood. Such Hancock Park beauties included the "El Royale," a classy "New York-style Spanish-French Revival" dwelling once home to Clark Gable, Loretta Young, and gangster film noir heavyweight George Raft. The Mauretania Apartments, a sleek eggshell-white structure, was commissioned and had its penthouse bought by Jack Haley, most remembered for his performance as the Tin Man and Hickory in the *Wizard of Oz*. The glamorous penthouse, which came complete with a ballroom and two fully furnished dining halls, was so dazzling that it was once chosen as a place of accommodation for President John F. Kennedy.

A picture of wildflowers and tar on the site

Notwithstanding Denton's best efforts, the tar pits' ties to prehistory – more specifically, the Pleistocene Period – remained unappreciated until 1901, when another well-respected petroleum geologist and former superintendent of the Union Oil Company's San Joaquin Valley branch named William "Bill" Warren Orcutt was sent to the ranch to reevaluate prospects for oil. Born in Minnesota's Dodge County, Orcutt was regarded as a trailblazer in the union of geology and the oil industry. It was, in fact, the Union Oil Company, then Unocal, that started the trend of installing a department of geology within commercial oil firms.

Whether or not the Stanford-educated geologist had ever perused Denton's work is unknown, but he immediately became obsessed with the La Brea Tar Pits on the very first day of the job. A hash of cream-colored bones that had emerged, caught in the underlying asphalt following the boring of a water well, swiftly snagged his attention. Orcutt rummaged through his tools, and though it did not contain the rock hammer, chisels, probes, and other instruments normally found in a paleontologist's field kit, the scrappy scientist succeeded in carving out a chunk of the hardened bone stew and sent it to the laboratory for further testing. The remnants were later determined to be a slab of the thick, osteoderm-studded (bony, usually circular bumps or plates embedded into the skin) pelt of an extinct ground sloth (or giant sloth), either the Nothrotheriops, Megalonyx, or the Paramylodon.

Orcutt, who was understandably giddy with excitement when he realized the historical goldmine he had uncovered, hastened over to the Hancock family and requested the green light to dig for more fossils of the like. The patch of land housing the water well that precipitated Orcutt's marvelous find was later renamed "Orcutt Field," and the oil well drilled here shortly later continues to generate copious amounts of oil to this day.

As thrilled as Orcutt was when he received excavation permit, the work the project required was as onerous as it was tedious. Pinpointing the locations of these bones was, interestingly enough, considerably easier in comparison to the actual act of retrieving them. Meticulously and consistently chipping away at the asphalt surrounding the precious specimen was a painfully mindnumbing and physically exacting process - it took, on average, an entire day's work to recover out a single specimen.

Over the next four years, Orcutt and his partner F. M. Anderson oversaw the paleontology digs at the ranch and continued to expand their quickly growing collection of prehistoric bones and specimen, among which included an extremely rare, complete skull of a saber-tooth tiger, as well as the remains of the extinct dire wolf. The final year of their excavation lease culminated with the pair uncovering a substantial hunk of hardened asphalt peppered with a mesmerizing miscellany of Pleistocene fossils – the largest of its kind ever discovered in a single setting. In April 1904, Unocal tasked Orcutt with designing and mapping out a new town site and suggested that the settlement be christened after the scientist. Orcutt, however, politely declined the offer, as he found the entire notion uncouth, comparing the act to "naming cheap cigars after cheap actresses." The paleontology world's decision to style the La Brea Coyote "canis latrans orcutti" as an homage to the geologist-turned-paleontologist was far more well-received.

A depiction of life around the tar pits

The Pleistocene Bones

"...We are a pack of dire wolves

Who scented death and mired ourselves...

Lion and lioness salivate

At bison ready trapped to eat.

Coyote, jaguar, and puma

Die for a taste of dying llama..." - Duncan Forbes, "La Brea"

In the winter of 1905, Orcutt turned to Stanford University and solicited from the institution funds to further finance his fossil exploits at the ranch. When his alma mater rebuffed his offer and instead advised him to try his luck with UC-Berkeley, this put him in touch with Dr. John Campbell Merriam, a fellow fossil connoisseur and the university's resident vertebrate paleontologist. Merriam, who became instantaneously enamored with Orcutt's astounding fossil collection, began to foster what soon became a solid friendship with him through frequent exchanges of letters. On behalf of UC-Berkeley, Merriam then requested and acquired permission from the Hancock family to perform an extensive scientific study in Rancho La Brea. His riveting article, "Death Trap of the Ages," which he published in *Sunset Magazine* in 1908, propelled the ranch's tar pits into the mainstream.

Merriam

The next 10 years marked the golden age of excavations at the La Brea Tar Pits. Dr. Merriam and his initially slight squad trawled their toes to test the depths of the paleontological potential in the tar pits as they awaited the needed funding for the sweeping excavation he had in mind. In the meantime, another fossil aficionado took the reins. In September 1907, a zoology and biology teacher employed at Los Angeles High School, James Zacchaeus Gilbert, chaperoned his first field trip to Rancho La Brea to catch a glimpse and whiff of the La Brea Tar Pits. That morning, Gilbert and his students filed into the community trolley, which took them to the fringes of the town, and disembarked at the terminal station.

Given the absence of present-day bus and subway stations, the party would have had to slog through quite a stretch before arriving at the ranch, but it was, as they would soon see, well worth the hike. Gilbert, a proponent of the kinesthetic method (hands-on learning), had arranged for his students to conduct their own amateur paleontological dig at the tar pits. Whether or not Gilbert had paid a fee for this privilege is uncertain, but the party carried on with their exhumation all the same.

The students split up into different crews and ferreted around the oozing springs of pitch aimlessly with magnifying glasses and probes until one of them zeroed in on a shard of bone protruding from the surface of one of the viscous pools. Thrilled by their good fortune, Gilbert and his students proceeded to hammer around the piece of bone, but they were unable to complete the task before it was time to catch the last trolley. As such, they concealed the bone with a heap of leaves and returned the following week to pick up where they had left off. The work eventually resulted in the disinterment of an extinct American lion.

Their propitious find kindled a passion that would soon consume Gilbert's life. In 1909, Gilbert became one of the first to captain the first large-scale excavation project at Rancho La Brea, spotlighting a 23-acre swatch of land encompassing Hancock Park. The clever professor and paleontology enthusiast brilliantly killed two birds with one stone by recruiting his students for the labor of the exhumations. Keen to expedite the paleontological excavations at the ranch, Gilbert was among the first to organize fundraisers and events that would fan the flames of interest surrounding the tar pits and their hidden jewels, and he managed to procure a steady stream of financial support from the Southern California Academy of Sciences and the Los Angeles County Board of Supervisors. The following year, Gilbert was selected to front a sizable project financed by the academy, appropriately named the "Academy Pit."

All in all, Gilbert and the Academy Pit crew succeeded in unearthing thousands of individual specimens and bones wedged dozens of feet underground. They located several dire wolves, giant ground sloths, mastodons, mammoths, and saber-toothed tigers, one of which was given the charming sobriquet "Smiley," among other prehistoric beasts. The ground sloth uncovered by Gilbert's crew was one of the rarest finds in the history of the ranch, one of only six ever found in the pits. A healthy percentage of the team's finds consisted of the carcasses of smaller mammals, which included that of coyotes, raccoons, and skunks. Gilbert also distinguished himself from the rest of the excavators, who were in hot pursuit of more substantial Pleistocene animals, by accumulating, rather than disregarding the tiny and brittle remnants of birds.

It seems Gilbert's obsession with the La Brea fossils was purely driven out of sincere interest and unfeigned curiosity, as opposed to wealth and international recognition. He received no compensation during the two years of the academy's excavations at the ranch, and in 1911, Gilbert, by then something of a household name in the Californian scientific community, became intimately involved in the founding and development of the Los Angeles County Museum of History, Science, and Art (now the Natural History Museum of Los Angeles County) at Exposition Park. The better part of Gilbert's fossil vertebrate treasury was donated to the museum, completed two years later, forming the core collection of the nascent establishment.

The skeleton of a Columbian mammoth found in the tar pits

Following the conclusion of Gilbert's time at the ranch in the early months of 1911, a separate unidentified excavation crew chanced upon another hoard of archaic treasures. Roughly 600 feet northwest of the old Hancock asphalt quarry, the diggers unearthed a staggering assemblage of compactly packed animal and bird remnants, interspersed with branches and twigs. It was a fortuitous find because it was outside of the grid sketched by the present excavation teams, nestled in a patch of land that had been shattered by dynamite sticks planted by Hancock's laborers several decades prior in a bid to see if any salable asphaltum lay underneath the surface.

Scientists were especially delighted by the finds, for the bones were encapsulated in a relatively soft, cushion-like matrix of tar, gravel, and sand, which meant that the fossils could be extricated with little effort. This lowered the risk of damaging the specimen. William Weston, author of *La Brea Tar Pits: An Introductory History*, provided a concise description of the "bottle-shaped" dig site, now referred to as "University of California Locality 2050," writing, "The topmost part of the pocket, or the neck of the 'bottle,' was about five feet wide. Below the neck, the pocket extended outward with increasing depth until at 10 to 12 feet it was about eight feet wide. The last remaining bones of the pocket were at a depth of 17 feet. The boundary forming the contour of the bottle had a lumpy irregularity as the pressurized tar had pushed its way unevenly in the surrounding green and brown clays."

In 1912, Professor Merriam finally secured the capital required for a large-scale excavation to rival that of Gilbert's. Like Gilbert, Merriam's excavation crews predominantly consisted of his students, past and present. Over the years, Merriam and his students – his most notable subordinate being Chester Stock of the California Institute of Technology – composed and recorded reams of descriptions regarding the fossils they recovered from their pits, which included the ancestors of now-common animals such as camels, peccaries, bears, and wolves. Merriam's team also retrieved the remnants of a saber-toothed cat, and in time the *Smilodon californicus* was ordained the official fossil of the state of California. A black-and-white rendering of its skull is used as the centerpiece of the Southern California Academy of Science's official crest.

After some time, Merriam was burdened with the hunt for a suitable home for his burgeoning collection. Along with the vast number of specimens, which were steadily growing and therefore at risk of misplacement and damage, the fossils reeked of asphalt. A decision was later made to transfer Merriam's fossils to Sather Tower, a 307-foot chiffon-white, slender clock and bell tower hybrid on the campus of UC-Berkeley, also colloquially referred to as "The Campanile," which is now home to approximately 300,000 fossils spread out across five floors. During this time, Merriam also exerted plenty of effort into building up and fleshing out the university's then-fledgling paleontology program in the hopes that it would one day reign as the best of its kind in the nation.

In the autumn of 1912, about a few months into the UC-Berkeley excavations piloted by Merriam, one of his crews came across yet another breathtaking find in the form of UC Locality 2051. Situated about 70 feet southeast from the Locality 2050, the crew found not one, but three separate compartments of fossils. The first stratum measured about 15 feet across and boasted a depth of about 22 feet, while the second and third layers lay eastward with depths of 21 feet and 14 feet, respectively. Much to the elation of the excavators, these fossils were also encased in a soft matrix of pitch and gravel and capped with a rigid layer of asphalt.

A 1913 bulletin authored by R.C. Stoner, entitled "Recent Observations on Mode of Accumulation of the Pleistocene Bone Deposits of Rancho La Brea," marveled at the strange nature of the bones' arrangement in this lot, which, while buried deep within the earth, were found densely packed together in fairly narrow spaces. A fragment from Stoner's report noted that "the pools [of pitch] were evidently large enough to catch one or two tigers, several wolves, and an ungulate at the same time, the latter serving as prey for the carnivores. This association is quite clearly shown in some places, and at one point in particular there were eight wolf skulls and many wolf bones mixed with the bones and skull of a large bison."

The pitch of the tar pits provided some of nature's best preservatives. Indeed, scrubbing the fossils clean with heated kerosene was a perilous and arduous process, but the black, viscous substance acted as an effective sealant that enveloped these bone bits and protected them against decay and degeneration, keeping them in near mint condition. This organic preservation method allowed the specimen to retain the subtlest and most delicate markings and details, such as the minuscule notches on the tooth of a carnivore, as well as imprints of networks of nerves and blood vessels. Even the full antennae, wings, and limbs of thumb-sized insects were occasionally discovered intact, and some blowfly pupae and other insect eggs were still clinging to bone marrow fissures. Without the La Brea-brand tar, it would have been extremely difficult for paleontologists to envision and establish grounded theories about southern California in the Ice Age.

Of course, regardless of the tar's exceptional preservative properties, the act of assembling the bones to form a complete creature – not unlike a free-form, frame-less puzzle – was an exceedingly intricate and toilsome task. Bones were strewn about and often entangled with the remnants of other creatures, and it was a rare stroke of luck to find a corpse of a tiny insect loosely strung together in the same place. The bones of the Pleistocene beasts were also severely mutilated, ranging from cracked fragments to saw-edged, splintered bits, which made it all the more difficult for scientists to piece together a coherent and cohesive skeleton in full.

These sponsored excavations at the ranch continued intermittently until 1913, and by then, Merriam and his diggers had exhumed thousands of bones from the tar pits. To facilitate the skeleton assembly process, Merriam traded similar fragments and specimens, which were classified as "duplicates," with other paleontologists across the world in return for components that they could afford to spare. In later years, Merriam partnered with his former student and current colleague, Chester Stock, once again, and the pair conducted an exhaustive case study centered on the *Smilodon californicus*. Their remarkable findings were later published in their 1932 treatise, *The Felidae of Rancho La Brea*, which profiled saber-toothed tigers and other extinct felines. The popular book was colored with exquisite drawings rendered by John Livsy Ridgway, a highly sought-after scientific artist who specialized in ornithological and paleontological artwork. Ridgway's trademark shadowing technique contributed to the dainty details and realism of his subjects. Stock, one of Ridgway's most ardent fans, raved about the illustrator's ability "to combine beauty with scientific accuracy in the delineation of fact."

Merriam is often credited with being one of the most vocal advocates of the proper preservation of the tar pit-derived fossils. Indeed, he later campaigned for the National Park Service to institute what he referred to as "a super university of nature," which he hoped would further endear the masses to Mother Nature's gifts. Stephen R. Mark, author of *John C. Merriam's Legacy in the State and National Parks*, explained, "Millions of visitors are indebted to Merriam in one form or another for their introduction to and understanding of geology, botany, plant succession, stream-side hydrology, fire regimes, and other fundamentals of field ecology." At the same time, Merriam continued to maintain the public's growing interest in the La Brea Tar Pits and the skeletal keys to the ancient past by penning more compelling articles that were published in various newspapers, magazines, and other publications.

Merriam's La Brea Tar Pits excavations played an instrumental role in fortifying his views on the theory of evolution. The following passage from Merriam's 1943 title, *The Garment of God*, explained, "But there is reason to believe that of concepts in science arising from study of nature, there are none that would be considered to have influenced our belief more deeply than the generalized principles concerning...evolution, reaching through vast ages in the story of the earth, and leading ultimately to advance in human life and institutions...As a result of this situation, one notes that in studying the universe widely in space, and deeply in time, out of our developing experience there tends to grow an attitude toward life that gives perspective instead of formless space, order in the place of aimless movement, confidence in the dependability of the universe and its laws, and faith that the world is so constructed as to maintain the trend of its...evolutionary progress. Such an attitude towards this world and its meaning is enormously important to us when, as now, complicated dangers and evils seem to almost overwhelm us..."

Stock also served as a crucial component of the whole operation. In 1925, Stock published a treatise on the ground sloths that once prowled La Brea. Five years later, he authored *A Record of Pleistocene Life in California,* then the first scientific book released by the Los Angeles County Museum of History, Science, and Art. His unforgettable experiences at the ranch prompted him to take his paleontology career to the next level, and over the next few years, he either headed or was involved in several other excavations in areas stretching from the southwest to northern Mexico, amassing an extraordinary private fossil library of his own. In later years, Stock co-founded the Department of Geology at Cal Tech in neighboring Pasadena, and he was awarded the title of Chief Curator of Science at the Los Angeles County Museum. In 1957, seven years after Stock's death, the entirety of his private collection, which included several specimens extracted from the La Brea Tar Pits, were acquired in bulk by the museum.

George Allan Hancock, the final owner of Rancho La Brea, was thrilled by the sudden explosion of scientific interest in the tar pits, but he began to fear that the waiting list of excavations from competing firms, institutions, and private parties would lead to the confused dispersal or disappearance of the pits' irreplaceable fossils. He was well-aware of the importance of coalescing these collections for better safekeeping, perhaps inspired by his father, who, under the recommendation of Denton, donated his library of Pleistocene mammal bones to the Boston Society of Natural History back in 1875. In 1913, Hancock declined to renew the contracts of the excavators and instead signed off on a strict lease with Los Angeles County, granting them the exclusive right to conduct digs in the ranch for a period of 24 months.

Between 1913 and 1915, county's excavation crews, overseen by a man named L.E. Wyman, proceeded to bore a total of 96 gaping holes in the ground with the sole objective of uncovering more fossil-rich pits with malleable matrices. Due to the lack of sufficiently advanced ground-penetrating radar technology, the excavators could not know the precise coordinates of these fossils, and they had no other choice but to rely on guesswork. They tore into the hard, exterior layer of asphaltum with heavy hammers, shovels, picks, and stuck sticks of dynamite onto unusually stubborn surfaces. They explored both active and inactive asphalt pools and navigated their way through the subterranean via man-made tunnels that sliced through rock and asphalt formations.

Although a substantial portion of the existing libraries of La Brea fossils today were exhumed between 1913 and 1915, which numbered over 750,000 individual specimens of flora and fauna altogether, the safety hazards of the dicey working environment was a cause for concern. Digging tools were severely antiquated in comparison to the instruments today. Shores and stilts were poorly built and unreliable, and the inefficiently built and poorly reinforced tunnels were prone to disastrous floods and were at risk of caving in at any given moment. The hot kerosene used to cleanse the remnants often caught fire and emitted toxic fumes. Worse yet, the laborers were pocketing a measly $3.50 (roughly $75 today) each day.

The 96 excavation sites were equal parts hits and misses. Among the eight major excavation sites were Pits 3, 4, 9, 61, 67, and 91. Like the bones disinterred by local universities, the remnants dug up by the county were in miserable shape, and a few significant pieces were disfigured due to their adjacency to water-soaked twigs and branches. An unnamed excavator who worked on Pit 4 cited an example of the damage in his report: "The disposition of this brush and the associated material as well as the markings on the brush itself indicate that this stuff was all washed in."

The eight major pits bore varying dimensions. The average pit was shaped like an upstanding funnel, its opening measuring about 15 feet in width, but these pits narrowed with depth, narrowing down a drop of roughly 25 feet to just a few inches in diameter. Weston explained, "These sediments form the outwash plain between the Santa Monica Mountains and the Pacific Ocean. Below this strata, the tar pit vent continues down through another layer of gravel, sand, and fine-grained marine sediments to oil reservoirs about 2,000 to 6,000 feet below the surface of the earth. The oil-bearing second layer is called [the] 'Upper Miocene,' and it forms the basin of the Los Angeles region."

Tar, sediment, and bones from varying time periods were often found in the same pit. The remnants discovered on the southern layers of Pit 9, for instance, were estimated to be around 38,000 years old, whereas the bones in the northern layers were no older than 13,500. Approximately 650 different species of flora and fauna have been unearthed from the La Brea sludge. Of the 650, 231 were vertebrates, 234 were invertebrates, and 159 classed as flora.

The rambling list of Pleistocene creatures and critters was a dizzying sight. Included in the lengthy catalog were remnants of mammoths, mastodons, giant ground sloths, American lions with tails measuring four feet in length, ancient bison, western horses, llamas, dwarf pronghorns, and hulking short-faced bears, the largest carnivorous animal that roamed North America during the Ice Age, which towered over surrounding animals with an 11-foot stature when on their hind legs. It also listed the distant, but recognizable cousins of common modern-day animals such as ancient dogs, pumas, jaguars, bobcats, gray foxes, raccoons, weasels, skunks, and field rodents. Two extinct species of camels were also discovered at La Brea, one of them measuring seven feet in height. Gopher snakes, garter snakes, western rattlesnakes, western pond turtles, frogs, rainbow trout, mollusks, clams, lizards, and other reptiles, as well as a slew of insects, such as pill bugs, termites, grasshoppers, snails, and millipedes, were just some of the names found in the mix. A distinct section was set aside for the various birds discovered in the pits, which included eagles, hawks, teratorn, vultures, and songbirds.

A skeleton of a short-faced bear found in the La Brea Tar Pits

Many of the county scientists' discoveries during this period altered a number of generally accepted views within the community. Larisa DeSantis, a paleontologist from Vanderbilt University, broke down one such revelation: "Isotopes from the bones previously suggested that the diets of saber-toothed cats and dire wolves overlapped completely, but the isotopes from their teeth give a very different picture. The cats, including saber-toothed cats, American lions, and cougars, hunted prey that preferred forests, while it was the dire wolves that seemed to specialize on open-country feeders like bison and horses. While there may have been some overlap in what the dominant predators fed on, cats and dogs largely hunted differently from one another."

Remains of Pleistocene plants also put the spotlight on the ecology of the Los Angeles Basin during the thaw of the Ice Age. As gathered by researchers from the Los Angeles County Museum of History, Science, and Art, the canyon redwood groves and sage scrubs endemic to the basin during the Last Glacial Period indicates a far moister, peninsula-like climate than previously thought. In the immediate decades that followed, lasting well into the late 1950s, scores of academics and scientific illustrators were employed to not only keep track of the ever-growing inventory, but also to produce monographs and composite sketches of the various new species of mammals, birds, and plants brought to light.

Over 4,000 specimens belonging to the dire wolf have been discovered in the pits thus far, making it the most recurrent of all the species uncovered within the ranch. Saber-toothed cats, of which there are 2,000 specimens, was ranked the second most common animal found in the tar pits, and coyote specimens, which approached 1,000, placed third. The oldest specimen in the database, an estimated 44,000 years in age, belonged to a dire wolf.

The most frequently discovered herbivores in the ranch was most likely the *Bison antiquus*, or the ancient bison, which were muscular creatures with an average shoulder height of seven and-a-half feet. This, in turn, suggests that Ice Age bison were, as dictated by nature, swamp animals, as opposed to their modern counterparts, who are classed as prairie animals.

More intriguing yet, a whopping 90% of the La Brea mammals and vertebrates trapped in the pitch were carnivores, meaning there were nine carnivores to every herbivore. This is a particularly headscratching tidbit considering that this is an inverse to the standard ratio. Based on two separate studies conducted in the mid-20th century, which examined the wolf-to-deer and lion-to-plant-eater populations in Ontario and Minnesota, as well as across Africa, herbivores, on average, outnumbered carnivores between 100-150:1. The discrepancy of the missing herbivores at La Brea, some experts believe, lends further credence to the tar entrapment theory. The plights of herbivores who found themselves immobilized by the glutinous asphalt were compounded by vicious and hungry carnivorous predators who became ensnarled in the sludge themselves, which potentially explained the excessive amount of carnivores in the pits.

Although people tend to think of dinosaurs when discussing fossils, not a single dinosaur has ever been discovered in the La Brea Tar Pits. The closest connection to dinosaurs were the 163 species of birds embedded in the asphalt, as avians are technically dinosaurs. La Brea's fossils, for the most part, ranged between 11,000-50,000 years in age, roughly 65 million years after the extinction of the last known dinosaur. The puzzling ratio of land birds to aquatic avians is yet another cold case that even scientists today have yet to close. Of the 163 species of birds discovered in the pits, only a trifling 8% were aquatic birds, namely geese, ducks, and other avians typically attracted to bodies of water. Much to the surprise of baffled scientists, the most commonly discovered bird was the turkey.

The most enjoyable part of the grueling aftermath was perhaps the construction of the narratives of the bones found in the pits. After all, each set of bones told its own story. For example, an admirably resilient timber wolf suffered a horrific amputation but managed to adjust to his condition for years before falling victim to the sludge. John M. Harris, editor of *La Brea and Beyond: The Paleontology of Asphalt-Preserved Biotas*, explained, "Two separate possibilities are proposed to explain the observed pathological condition: (1) a [false joint] resulted from osseous nonunion, where bone ends remained separated by soft tissue in life, or (2) a complete amputation event occurred with subsequent healing of the remaining bony shaft. Either case indicates that the animal survived for a significant amount of time – likely years – after the traumatic injury was inflicted."

The bones of ancient gray wolves, cougars, and coyotes, when juxtaposed with the herculean skeletons of mammoths, mastodons, and short-faced bears, suggested that they were once considered small and meek, almost docile creatures. These comparatively petite predators tracked down and feasted on smaller critters, and they often hid themselves behind trees and boulders to pick at the corpses left behind by larger predators. DeSantis described the significance of this finding: "The...exciting thing about this research is we can actually look at the consequences of this extinction. The animals around today that we think of as apex predators in North America – cougars and wolves – were measly during the Pleistocene. So when the big predators went extinct, as did the large prey, these smaller animals were able to take advantage of that extinction and become dominant apex predators."

While the bones and fragmented remains of flora and fauna were a dime a dozen, the only human remnants ever extracted from the sludge belonged to that of the infamous "La Brea Woman." In 1914, one of the county's crews fished out a human skull, along with 17 other bones, forming a partial skeleton. Her gender was determined by the shape and size of her pelvis. According to the profile woven together by researchers, the unidentified woman, who died about 9,000 years ago, was roughly 4'10" in height and was anywhere between 18-24 years in age at the time of her death. Judging by the stains, as well as the wear and tear on what was left of her missing teeth, the woman subsisted on a diet of mostly stone-ground grains. Upon further examination of her skull, some scientists have concluded that she belonged to the Chumash tribe, but others are not prepared to take that leap and have instead suggested that she belonged to some kind of hunter-gatherer tribe from the Paleo-Indian Clovis Culture. Melissa Cooper, a former volunteer at the ranch museum, noted, "There are hints within the skull that she may have had Native American features. You can tell by the way the nose was pointed and the depth of her eyes. Based on the skull, she had Asian features which does coincide with Native Americans."

The most interesting feature of the La Brea Woman's skull, however, were the visible, man-made fractures. This hinted at a violent death, most likely blunt force trauma, making her the first known victim of homicide in Los Angeles. Fragments of a stone mortar, often used in the burials of Southern Californian Native Americans, as well as the bones of a tamed dog, were also discovered just a few feet away from her remains.

A replica of her skull

A Legacy for the Ages

"...Fearful is the chance of poison:

Fearful, too, the great unknown:

Magic brings some positivists

Humbly on their marrow bone..." - Charles Kingsley, "The Legend of La Brea"

Three years after the secretive exhumation of the La Brea Woman, the ranch received a visitor named George C. Page, a young man who had recently relocated to California from Nebraska. Page had placed Rancho La Brea at the top of his itinerary following the early paleontology craze, but he was saddened when he realized that the Pleistocene skeletons he had been so desperate to scrutinize up close "were not on-site, but seven miles away at the Natural History Museum." This, Page promised himself, would soon change.

As the years progressed, Hancock became growingly invested in the preservation of the La Brea fossils, as well as the conservation of the tar pits and the overall property. In 1924, he donated 23 acres of the Rancho La Brea to Los Angeles County, an enclosure that thenceforth became known as "Hancock Park." The generous donation came with the proviso that the area be regarded as a "protected park," and that a museum housing the Pleistocene fossils uncovered from the tar pits be erected. Due to the construction of Hancock Park, paleontological digs within the ranch greatly declined in their frequency, at least for the next 45 years.

In the meantime, references to the La Brea Tar Pits began to multiply in the local papers for reasons that were anything but scientific. These stories, however, only amplified the buzz surrounding the mysterious sludge. On October 25, 1935, 18-year-old Mary Alice Bernard vanished seemingly without a trace, but not before leaving her mother with a cryptic message that told her, "Some day, you'll find me in the bottom of the La Brea pits." A passage from an article in the October 26th issue of the *San Bernardino Sun*, sensationally titled "Police Hunt Girl in Ancient Tar Pits at L.A.," told readers, "Police tonight searched with iron grappling hooks through ancient pits, more than 1,000,000 years old for...18-year-old...Bernard...Mrs. Minnie C. Bernard, the girl's mother, appealed to police to search the pits. The girl left home Tuesday night to mail a letter at a mailbox near the pits, and it was feared she may have fallen [or leapt] in. First efforts to drag the mire were balked by lightness of grappling equipment. Fire hooks, propelled downward by heavier weights, were brought to the pits."

The police hovered over the tar pits for hours on end in vain, but this perplexing tale took a sharp turn four months later when the grief-stricken mother spotted an all-too-familiar face in the paper. It was a dazed Mary Alice sporting a hospital gown, identified as an "amnesiac victim" who claimed to bear the name of "Ann Page." Mrs. Bernard sped to the hospital in downtown Los Angeles the very next morning, and a tearful reunion ensued. Whether or not "Ann Page" was actually Mary Alice, a misguided amnesiac, or a heartless impostor remains uncertain.

The La Brea Tar Pits appeared in the newspaper again just three years later after police and firemen were once again dispatched to the pits in December 1938 when passersby discovered a heartrendingly-worded suicide note signed by a George W. Page (no relation), as well as a pile of men's clothes just inches away from an asphalt spring. Imagine the astonishment and annoyance of the law enforcers when none other than the victim himself came up to the search site, completely unharmed, and sheepishly recounted his wildly ill-conceived plan. As it turned out, he had staged his own suicide in an effort to avoid his girlfriend. Page was subsequently arrested and sentenced to 30 days in the county jail for wasting government resources.

In 1963, the La Brea Tar Pits earned a designation as a National Natural Landmark by the National Parks Service. On June 13, 1969, once Hancock Park had taken shape, the "Rancho La Brea Project" was launched, which aimed to resume the excavation of the 10-foot-wide Pit 91 that was originally cracked open in 1915. This day is now commemorated as "Asphalt Friday." By the 1960s, excavation tactics and objectives had evolved, and collecting biases had been retired. The second round pinned its focus on harvesting every single fossil discovered in the pit, as opposed to only those of sizable mammals and vertebrates, and it dealt mainly with the excavations of smaller reptiles, amphibians, birds, insects, plants, and mollusks.

By the 1970s, George C. Page had established himself as a real estate tycoon and conqueror of various industries, and he was one of the most affluent men in the county. In 1975, Page made good on his promise and singlehandedly financed the entire construction of the museum that had been stipulated in the Hancock Park grant. The George C. Page Museum of La Brea Discoveries made its public debut for the first time in 1977. When construction workers drilled into the site of the museum to lay down the structure's foundations, they lucked into what was then the largest and most diverse deposit of fossils ever discovered at one time on the premises. 20 massive blocks of flora and fauna remnants encrusted in solidified matrices were wrenched out of the deposit.

In 2006, the ongoing excavation of the 1969 Rancho La Brea Project screeched to a halt due to the impromptu launching of Project 23. Before the crews commissioned by the Los Angeles County Museum of Art could begin their work on the foundations of a new subterranean parking lot, a salvage archaeologist named Robin Turner was tasked with inspecting the excavation site, which was roughly 9,290 square miles in size. The decision to appoint the specialist, as it turned out, was a wise one, because 16 deposits brimming with fossils and ancient relics were discovered, including 80% of the skeleton of a Columbian mammoth named "Zed."

About three months later, 23 wooden crates stocked with these plastic-swaddled deposits, which weighed up to 125,000 pounds, were lugged out of the ground with cranes. They were then bussed over to the central research facility of the Page Museum. The laboratory, otherwise known by visitors as the "Fish Bowl," was fitted with glass walls so that spectators could observe the paleontologists at work in real time.

The following year, researchers from UC-Riverside published a report concluding that the tar pits' bubbles were caused by "hardy forms" of naturally-occurring bacteria found in raw asphalt. When asphalt interacts with petroleum, these hosts of bacteria are activated, releasing the methane that engenders the bubbles in question. Around 200-300 species of new bacteria have since been identified in the La Brea Tar Pits.

In 2013, LAPD Sergeant David Mascarenas became the first known person to voluntarily plunge into the tar pits. Mascarenas, as reported by various news outlets, dove 17 feet into the gooey abyss to retrieve weapons and other evidence pertaining to a cold homicide investigation that had been discarded in the pits. In an interview with the *Los Angeles Times*, he said, "I've been under moving ships, [and] in underwater reservoir sheds. This is by far the craziest thing I've ever done...Visibility was zero. I could pretty much not see my hand until I put it up to my face."

In early 2019, the George C. Page Museum and Tar Pits was collectively reintroduced to the public as simply "The La Brea Tar Pits and Museum." All in all, over 3.5 million (in some sources, as many as six million) individual fossils have been extracted from the pits to date. The La Brea Tar Pits remains the "only active urban paleontology site in the world," with teams of paleontologists busy conducting ongoing excavations 361 days of the year, with the exceptions of the Fourth of July, Thanksgiving, Christmas, and New Year's Day.

A picture of the museum

In the summer of 2019, authorities from the La Brea Tar Pits and Museum unveiled a new project that aimed to study the tar pits during the Holocene Period, which could allow scholars to better understand how and why the Pleistocene creatures were ushered out of existence by mankind.

Online Resources

Other books about ancient history by Charles River Editors

Other books about La Brea on Amazon

Bibliography

Berko, L. (2013, October 28). A Century of Finding Awesome Dead Things at the La Brea Tar Pits. Retrieved December 1, 2019, from https://www.vice.com/en_us/article/mgbqp8/a-century-of-finding-awesome-dead-things-at-the-la-brea-tar-pits.

Bertao, D. E., & Dias, E. M. (1987, September). "El Portuguese": Don Antonio Rocha California's First Portuguese Sailor. Retrieved December 1, 2019, from https://www.jstor.org/stable/25158437?seq=1.

Cooper, A. (2010, May 27). STICKY SITUATION AT THE TAR PITS. Retrieved December 1, 2019, from https://www.laweekly.com/sticky-situation-at-the-tar-pits/.

Curwen, T. (2019, May 9). Tongva, Los Angeles' first language, opens the door to a forgotten time and place. Retrieved December 1, 2019, from https://www.latimes.com/projects/la-me-col1-tongva-language-native-american-tribe/.

Dell'amore, C. (2011, October 7). Tar Shrank Heads of Prehistoric Californians Over Time? Retrieved December 1, 2019, from https://www.nationalgeographic.com/news/2011/10/111006-tar-toxic-pollution-chumash-health-indians-science-heads/.

Editors, B. E. (2017). Henry Hancock. Retrieved December 1, 2019, from https://eng.lacity.org/henry-hancock.

Editors, C. R. (2002). LA BREA TAR PITS: AN INTRODUCTORY HISTORY (1769–1969). Retrieved December 1, 2019, from https://creationresearch.org/labrea-5/.

Editors, C. L. (2016, October 13). Amnesiacs, hoaxes, and homicide cover-ups at the La Brea Tar Pits. Retrieved December 1, 2019, from http://creepyla.com/2016/10/13/amnesiacs-hoaxes-and-homicide-cover-ups-at-the-la-brea-tar-pits/.

Editors, E. C. (2019, October 30). Denton, William (1823-1883). Retrieved December 1, 2019, from https://www.encyclopedia.com/science/encyclopedias-almanacs-transcripts-and-maps/denton-william-1823-1883.

Editors, F. C. (2011). JUAN CRESPI. Retrieved December 1, 2019, from https://factcards.califa.org/exp/crespi.html.

Editors, F. W. (2011). *Los Angeles in the 1930s: The Wpa Guide to the City of Angels.* University of California Press.

Editors, G. H. (2013, June 10). June's SPIRITS with SPIRITS. Retrieved December 1, 2019, from http://ghoula.blogspot.com/2013/06/junes-spirits-with-spirits.html.

Editors, L. T. (2009, November 24). The skeleton that the Page Museum doesn't want you to see. Retrieved December 1, 2019, from https://latimesblogs.latimes.com/culturemonster/2009/11/the-skeleton-that-the-page-museum-doesnt-want-you-to-see.html.

Editors, L. A. (2017). The La Brea Woman. Retrieved December 1, 2019, from http://www.laalmanac.com/history/hi02v.php.

Editors, L. A. (2018). HISTORY TIMELINE Los Angeles County Pre-History to 1799 A.D. Retrieved December 1, 2019, from http://www.laalmanac.com/history/hi01a.php.

Editors, L. A. (2018). The Presence of the Past: Peter Zumthor Reconsiders LACMA Didactics . Retrieved December 1, 2019, from https://www.lacma.org/sites/default/files/The Presence of the Past-didactics.pdf.

Editors, N. G. (2015, May 14). Sadieville resident to receive honorary doctorate. Retrieved December 1, 2019, from http://www.news-graphic.com/news/sadieville-resident-to-receive-honorary-doctorate/article_27cdaede-f9b6-11e4-9975-039662647958.html.

Editors, N. H. (2019). La Brea Tar Pits History. Retrieved December 1, 2019, from https://tarpits.org/la-brea-tar-pits-history.

Editors, N. H. (2019). Two Stops One Ice Age. Retrieved December 1, 2019, from https://tarpits.org/stories/two-stops-one-ice-age.

Editors, N. H. (2019). Botany Collections. Retrieved December 1, 2019, from https://tarpits.org/research-collections/tar-pits-collections/botany-collections.

Editors, N. H. (2019). Early Excavations. Retrieved December 1, 2019, from **https://tarpits.org/early-excavations**.

Editors, N. H. (2019). Historic Research at Rancho La Brea. Retrieved December 1, 2019, from https://tarpits.org/historic-research-rancho-la-brea.

Editors, P. S. (2015). 1860: Henry Hancock and Rancho La Brea. Retrieved December 1, 2019, from http://www.playgroundtothestars.com/timeline/1860-henry-hancock-and-rancho-la-brea/.

Editors, S. I. (2015). LA BREA TAR PITS FACTS. Retrieved December 1, 2019, from https://someinterestingfacts.net/la-brea-tar-pits-facts/.

Editors, S. T. (2016, June 28). 15 Things You Never Knew About the La Brea Tar Pits. Retrieved December 1, 2019, from https://suburbanturmoil.com/about-the-la-brea-tar-pits/2016/06/28/.

Editors, S. C. (2016). Chumash History. Retrieved December 1, 2019, from https://www.santaynezchumash.org/history.html.

Editors, S. E. (2016). Yesterday - A Rich History. Retrieved December 1, 2019, from http://santamariaenergy.com/about-us/yesterday/.

Editors, S. M. (2017, October 1). G. Allan Hancock has a colorful history. Retrieved December 1, 2019, from https://santamariatimes.com/lifestyles/columnist/shirley_contreras/g-allan-hancock-has-a-colorful-history/article_b72077c5-f3fd-5746-8f73-1e46823aa581.html.

Editors, S. M. (2017, May 21). Heart of the Valley: The 'Dean of Petroleum Geology'. Retrieved December 1, 2019, from https://santamariatimes.com/lifestyles/columnist/shirley_contreras/heart-of-the-valley-the-dean-of-petroleum-geology/article_897992a4-5a0e-5304-a1b0-05b97a236f5e.html.

Editors, T. A. (2013). 7 Most Interesting Facts In La Brea Tar Pits History. Retrieved December 1, 2019, from https://traveladvisortips.com/7-most-interesting-facts-in-la-brea-tar-pits-history/.

Editors, U. B. (2009). John C. Merriam (1869-1945). Retrieved December 1, 2019, from https://ucmp.berkeley.edu/history/merriam.html.

Editors, U. B. (2012). Localities of the Pleistocene: The La Brea Tar Pits. Retrieved December 1, 2019, from https://ucmp.berkeley.edu/quaternary/labrea.html.

Editors, U. C. (2016). San Bernardino Sun, Volume 42, 26 October 1935. Retrieved December 1, 2019, from https://cdnc.ucr.edu/cgi-bin/cdnc?a=d&d=SBS19351026.1.1&e=-------en--20--1--txt-txIN--------1.

Editors, V. U. (2019, August 5). Intense look at La Brea Tar Pits explains why we have coyotes, not saber-toothed cats. Retrieved December 1, 2019, from https://phys.org/news/2019-08-intense-la-brea-tar-pits.html.

Editors, W. B. (2013, March). 3189 Wilshire Boulevard. Retrieved December 1, 2019, from https://wilshireboulevardhouses.blogspot.com/2013/03/3189-wilshire-boulevard.html.

Editors, W. (2018, December 16). Henry Hancock. Retrieved December 1, 2019, from https://en.wikipedia.org/wiki/Henry_Hancock.

Editors, W. (2018, June 20). William Warren Orcutt. Retrieved December 1, 2019, from https://en.wikipedia.org/wiki/William_Warren_Orcutt.

Editors, W. (2019, November 27). La Brea Tar Pits. Retrieved December 1, 2019, from https://en.wikipedia.org/wiki/La_Brea_Tar_Pits.

Editors, W. (2019, October 17). Rancho La Brea. Retrieved December 1, 2019, from https://en.wikipedia.org/wiki/Rancho_La_Brea.

Gannon, M. (2017, June 23). 10 Fascinating Facts About the La Brea Tar Pits. Retrieved December 1, 2019, from https://www.mentalfloss.com/article/501974/10-fascinating-facts-about-la-brea-tar-pits.

U.S. Government Printing Office. (1949). *Geological Survey Bulletin.*

Harris, J. M. (Ed.). (2015, September 15). La Brea and Beyond: The Paleontology of Asphalt-Preserved Biotas. Retrieved December 1, 2019, from https://tarpits.org/sites/default/files/2019-05/la_brea_and_beyond_2015._nhm_science_science_no._42.pdf.

Hull, K. (2019). Hancock Part: A Place Apart. Retrieved December 1, 2019, from http://www.discoverhollywood.com/Publications/Discover-Hollywood/2019/Discover-Hollywood-Fall-2019/Hancock-Part-A-Place-Apart.aspx.

Jackson, W. (2018). Lessons from the La Brea Tar Pits. Retrieved December 1, 2019, from https://www.christiancourier.com/articles/218-lessons-from-the-la-brea-tar-pits.

Kantor, L. (2017). History Trapped in the La Brea Tar Pits. Retrieved December 1, 2019, from https://www.splicetoday.com/writing/history-trapped-in-the-la-brea-tar-pits.

Khan, A. (2015, September 9). Great Read: Within UC Berkeley's famous tower, a scarcely known trove of fossils. Retrieved December 1, 2019, from https://www.latimes.com/science/great-reads/la-sci-c1-belltower-bones-20150909-story.html.

Kudler, A. G. (2013, June 7). 7 Descriptions of Diving in the La Brea Tar Pits From the Only Guy Who's Done It. Retrieved December 1, 2019, from https://la.curbed.com/2013/6/7/10236878/7-descriptions-of-diving-in-the-la-brea-tar-pits-from-the-only-guy.

Kyle, D. E. (2002). *Historic Spots in California: Fifth Edition.* Stanford University Press.

Lloyd, A. (2017, October 9). A Brief History Of L.A.'s Indigenous Tongva People. Retrieved December 1, 2019, from https://laist.com/2017/10/09/a_brief_history_of_the_tongva_people.php.

Lund, B. (1997). *The Chumash Indians.* Capstone.

McNassor, C. (2011). *Los Angeles's La Brea Tar Pits and Hancock Park.* Arcadia Publishing.

Meares, H. (2016, February 24). The Lady of La Brea: Madame Ida Hancock Ross, Los Angeles' Forgotten Matriarch. Retrieved December 1, 2019, from https://www.kcet.org/history-society/the-lady-of-la-brea-madame-ida-hancock-ross-los-angeles-forgotten-matriarch.

Meares, H. (2018, June 7). 13 glamorous apartments from Hollywood's Golden Age. Retrieved December 1, 2019, from https://la.curbed.com/2018/6/7/17382456/glamorous-apartments-for-rent-old-hollywood.

Morris, J. D., & Clarey, T. (2013, May 31). The La Brea Tar Pits Mystery. Retrieved December 1, 2019, from https://www.icr.org/article/la-brea-tar-pits-mystery.

Pauls, J. (2012, January 18). McPherson man made history at La Brea Tar Pits. Retrieved December 1, 2019, from https://www.mcphersonsentinel.com/article/20120118/NEWS/301189937.

Piper, R. (2009). *Extinct Animals: An Encyclopedia of Species that Have Disappeared during Human History: An Encyclopedia of Species that Have Disappeared during Human History.* ABC-CLIO.

Plax, A., & Diaz, C. (2017, September 22). Jaeger Museum awarded $10k NEA grant. Retrieved December 1, 2019, from https://lvcampustimes.org/2017/09/jaeger-museum-awarded-10k-nea-grant/.

Rosen2, J. (2014, June 21). A wrinkle in time: Finding the ice age in urban Los Angeles. Retrieved December 1, 2019, from https://www.latimes.com/science/sciencenow/la-sci-sn-la-brea-20140620-story.html.

Schultz, C. (2013, October 28). Animals Trapped in the La Brea Tar Pits Would Take Months to Sink. Retrieved December 1, 2019, from https://www.smithsonianmag.com/smart-news/animals-trapped-in-the-la-brea-tar-pits-would-take-months-to-sink-6006035/.

Shaw, E. (2015, January 5). Nature's Time Capsules: A Guide to the World's Pitch Lakes. Retrieved December 1, 2019, from https://www.atlasobscura.com/articles/nature-s-time-capsules-a-guide-to-the-world-s-pitch-lakes.

Spence, M. (2006). Southern California Quarterly. Retrieved December 1, 2019, from https://scq.ucpress.edu/content/87/4/418.

Spitzerri, P. R. (2017, January 26). Drilling for Black Gold: La Brea Oil Field, 1920s. Retrieved December 1, 2019, from https://homesteadmuseum.wordpress.com/2017/01/26/drilling-for-black-gold-la-brea-oil-field-1920s/.

Strickland, A. (2019, October 11). Preserving the unique history of the La Brea Tar Pits. Retrieved December 1, 2019, from https://edition.cnn.com/2019/10/11/world/la-brea-tar-pits-3d-scanning-scn/index.html.

Strickland, A. (2019, August 5). Fossils in La Brea Tar Pits reveal why coyotes still exist, but not saber-toothed cats. Retrieved December 1, 2019, from https://edition.cnn.com/2019/08/05/world/la-brea-tar-pits-coyotes-scn-trnd/index.html.

Switek, B. (2019). *The Secret Life of Bones: Their Origins, Evolution and Fate*. Prelude Books.

Thompson, E. (2010, September 25). Closed Mondays: George C. Page Museum. Retrieved December 1, 2019, from https://laist.com/2010/09/25/closed_mondays_george_c_page_museum.php.

Toothman, J. (2017). How the La Brea Tar Pits Work. Retrieved December 1, 2019, from https://science.howstuffworks.com/environmental/earth/archaeology/la-brea-tar-pits1.htm.

Trinidad, E. (2014, September 17). April 1977 - George C. Page Museum Opens at La Brea Tar Pits. Retrieved December 1, 2019, from https://www.kcet.org/kcet-50th-anniversary/april-1977-george-c-page-museum-opens-at-la-brea-tar-pits.

Twain, L. (2015, July 14). Rancho La Brea and The Tar Pits. Retrieved December 1, 2019, from https://news.thelandpatents.com/rancho-la-brea-and-the-tar-pits/.

Udurawane, V. (2016). Trapped in tar: The Ice Age animals of Rancho La Brea. Retrieved December 1, 2019, from http://www.eartharchives.org/articles/trapped-in-tar-the-ice-age-animals-of-rancho-la-brea/.

Walker, A. (2013, October 29). The La Brea Tar Pits Remind Us That Los Angeles is an Ancient City. Retrieved December 1, 2019, from https://gizmodo.com/the-la-brea-tar-pits-remind-us-that-los-angeles-is-an-a-1453499766.

Wells, B. (2019, July 29). Discovering the La Brea "Tar Pits." Retrieved December 1, 2019, from https://aoghs.org/energy-education-resources/discovering-oil-seeps-2/.

Weston, W. (2002, December). LA BREA TAR PITS: A CRITIQUE OF ANIMAL ENTRAPMENT THEORIES. Retrieved December 1, 2019, from https://creationresearch.org/la-brea-tar-pits-critique-animal-entrapment-theories/.

Wilson, S. (2011, March 11). 5 COOLEST NEW FOSSILS DISCOVERED IN THE LA BREA TAR PITS (PHOTOS). Retrieved December 1, 2019, from https://www.laweekly.com/5-coolest-new-fossils-discovered-in-the-la-brea-tar-pits-photos/.

Free Books by Charles River Editors

We have brand new titles available for free most days of the week. To see which of our titles are currently free, click on this link.

Discounted Books by Charles River Editors

We have titles at a discount price of just 99 cents everyday. To see which of our titles are currently 99 cents, click on this link.

Printed in Great Britain
by Amazon